THE
POCKET
Herbology

Published in 2025
by Gemini Books
Part of Gemini Books Group

Based in Woodbridge and London

Marine House, Tide Mill Way
Woodbridge, Suffolk IP12 1AP
United Kingdom
www.geminibooks.com

Text by Becky Freeth
Cover image: Shutterstock Ltd/ifiStudio

ISBN 978-1-80247-297-4

Manufacturer's EU Representative: Eurolink Compliance Limited, 25 Herbert
Place, Dublin, D02 AY86, Republic of Ireland. admin@eurolink-europe.ie.

Disclaimer: This book is intended for general informational purposes only and
should not be relied upon as recommending or promoting any specific practice
or method of health treatments. It is not intended to diagnose, treat or prevent
any illness or condition and is not a substitute for advice from a health care
professional. You should consult your health practitioner before engaging in any
of informational detailed in this book. You should not use the information in this
book as a substitute for health or other treatment prescribed by a professional
practitioner. The publisher makes no representations or warranties with respect
to the accuracy, completeness or currency of the contents of this work, and
specifically disclaim, without limitation, any implied warranties of merchantability
or fitness for a particular purpose and any injury, illness, damage, liability or loss
incurred, directly or indirectly, from the use or application of any of the contents of
this book. Furthermore, the publisher is not affiliated with and does not sponsor
or endorse any methods of treatment or products referred in this book.

Printed in China

10 9 8 7 6 5 4 3 2 1

MIX
Paper | Supporting
responsible forestry
FSC
www.fsc.org   FSC® C020056

THE
POCKET

Herbology

G:

CONTENTS

Introduction

Beyond our walls, there is a world of plants that bring therapeutic and medicinal value to humans and animals. Animals have naturally evolved to forage for (and avoid) certain species and, as science has improved, humans have developed a greater understanding of herbs that can heal, boost, soothe and improve the mind and body.

This relationship between people and plants has existed since the dawn of time, but only now can we put real evidence behind the theories that have long been the stuff of folklore.

Keep this pocket book handy as you discover the history of herbology and the mighty power of over 30 plants.

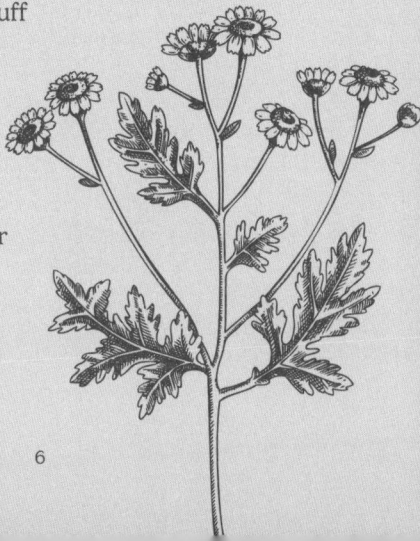

"It is through the alchemy of herbs that we not only heal our bodies, but also nourish our souls."

Rosalee de la Forêt,
Alchemy of Herbs (2017)

Chapter One

HERBOLOGY 101

"Earth has no sorrow that earth cannot heal."

John Muir,
John of the Mountains, 1872

Herbs for Healing

Can plants really heal a broken heart? Can they improve the weather, protect us from evil or help us live forever?

Before scientific study, it was easy to believe that plants had magical properties. Modern herbal medicine is less about wizardry and more about how humans have learned the myriad ways plants can help us.

Herbology

A herb can be any plant with leaves, seeds or flowers used for nutrition, flavouring, medicine or perfume.

Herbology is the study of plants, by botanists. They collect and study plants to identify their species and properties from a scientific perspective. Herbology also includes the study of plant lore, and is the foundation on which herbalism is built.

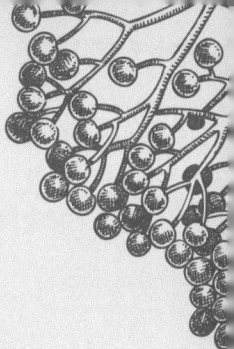

Herbalism

Herbalism is the art and practice of using herbs for their benefits in supporting health and wellbeing.

Herbalists will study the rich botanical history of plants, and explore their properties for remedial and medicinal use.

Leaves, roots, seeds and flowers are used fresh or dried to make teas, infusions, tinctures or salves. Herbs are also ground to powder to sprinkle on food or consume as capsules. There are specific herbalist qualifications in different countries; it's also easy to learn the basics of using herbs for your health and wellbeing – read on!

Herbals & Their Legacy

A "herball" is a book detailing the names and virtues of plants and herbs, the first of which was published in 1525. Later, these "plant bibles" featured elaborate illustrations and quickly gained popularity as handbooks of human discovery.

In 1597, English gardener John Gerard wrote the revolutionary *Generall Historie of Plantes* (commonly referred to as Gerard's *Herball*) featuring an exhaustive list of 1,039 rare plants and their properties.

By 1653, a physician with a combined knowledge of herbalism and astrology named Nicholas Culpeper published a ground-breaking manual called *The English Physician (Culpeper's Complete Herbal).*

It featured over 400 herbs, when and where to find them, and how to use them for healing. It is one of the most enduring herballs in history.

Another notable amateur botanist was John Evelyn – known to most as a historical diarist. He spent years painstakingly compiling an encyclopedic history of gardening.

Many traditional herballs are still available to purchase now – still with their striking original line drawings.

Families & Names

We may know a common herb like sweet basil by one traditional name, but botanists call it "Ocimum basilicum" because herbs are assigned to a family (such as Lamiaceae) and given a botanical name (often Latin) to help with identification.

Look out for these common identifiers:

1. **"X"**: One species has been cultivated with another, or they have grown together in the wild. For example, "*Mentha x piperita*" denotes the hybrid peppermint.

2. **VULGARE:** Indicates the common variety of the herb.

3. **OFFICINALIS:** A herb officially used by the ancient apothecaries.

4. **SPP:** Denotes that there are several species of this herb.

Foraging for Beginners

You are legally allowed to pick fruit, flowers, fungi and foliage from public land for personal use. Follow these simple tips for gathering herbs outside the home:

1. Take with you gardening gloves, a box for collection, sharp scissors or shears.

2. Stay on marked paths to protect crops and wildlife. Avoid protected areas. Seek the landowner's permission on private land.

3. Gather herbs during dry spells for better preservation.

4. To collect seeds, cut ripened seedheads lower down the stalk.

5. Take only what you need. Leave enough for wildlife and other foragers.

6. Leave plant roots alone: in many places it is unlawful to uproot a wild plant.

7. Stay away from vulnerable, rare or endangered species.

8. Leave the site as you found it.

Preserving Plants

Growing Herbs
You don't need a sprawling back garden to propagate new plants. Indoor environments are ideal for growing chillies and basil, and many herbs are hardy enough to survive in a pot outside year round.

For the green-fingered, a great beginner medicinal herb garden could include: basil, chamomile, oregano, peppermint, rosemary and sage.

Drying Herbs
You can, of course, buy dried herbs, but it is easy to dry your own:

- Gather fresh herbs into small bunches and hang them upside down, out of direct sunlight, for drying for around one month.

- Dry out seedheads in a bag until they are ready to be shaken out or gently separated.

- Dried herbs and seeds can then be stored inside airtight jars.

"There are no worthless herbs – only the lack of knowledge."

Ibn Sina (980–1037 CE)

Preparation of Herbs

It is growing more and more convenient to purchase essential oils, ointments, tea leaves, capsules and dried herbs.

Throughout this book, you will also find directions to help you create your own ointments, infusions and salves, using herbs that you have foraged or grown. Here are some methods you may come across:

DECOCTION: Made by extracting the herb's beneficial elements through boiling, decoctions can be gargled to help with sore throats.

ESSENTIAL OIL: You can buy the essential oil of many of the herbs. To use the oil, it must often be mixed with a carrier oil, like jojoba or coconut oil.

INFUSION: Herbs can be steeped in, or infused with, hot water to create herbal tea.

MOXIBUSTION: A method used in Traditional Chinese Medicine, involving burning leaves and waving the smoke around the body's energy points.

POULTICE: A poultice is created by crushing herbs into a pulp using a pestle and mortar, and adding just enough water to make a paste. It can then be used topically.

SUPPLEMENTS: Pills, tablets or capsules containing herbal extracts. Purchased at any good food store, and easy to add into your daily routine.

TINCTURE: A tincture is made up of plant extract dissolved in alcohol. It is then usually diluted in water and consumed.

Chapter Two

HISTORY & FOLKLORE

A Spiritual Significance

Throughout history – and still today in many cultures – herbs are valued with a high spiritual significance. Rituals, ceremonies and meditation practices often form part of their experience, whether for luck, health, magic, purification, abundance or love.

Traditionally, recipes, superstitions and stories were passed on to imbue the next generation with wisdom about their sacred connection.

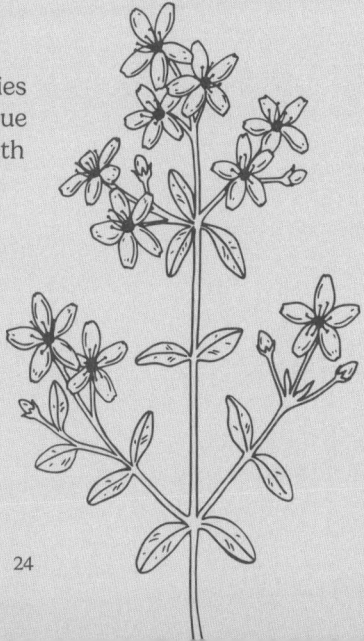

Mandrake: Myth or Magic?

A hallucinogenic perennial known as mandrake is the original "magic" herb believed to possess the power of dark earth spirits.

In medieval times, it was made into charms to bring love and luck, but humans were warned against ever uprooting it with their bare hands. Instead, it could only be pulled up in the moonlight, by a black dog attached by a cord. Even then, the uprooting should be preceded by ritual and prayer.

The mandrake even featured in J.K. Rowling's Harry Potter stories as a fatal root with a human face and a deathly scream.

Cleopatra's Beauty Secret

The enigmatic queen of Egypt is said to have maintained a keen interest in plants, collecting a long list of natural ingredients for personal use. She is said to have used an early technique called "enfleurage" to capture the essence of aromatic flowers like lavender for perfume.

She also knew the remarkable perks of aloe vera and smothered her skin in the gel to capture eternal youth.

Serpents & Mad Dogs

In *Culpeper's Complete Herbal*, Nicholas Culpeper highlighted no fewer than 20 herbs that could help with snake bites.

According to his herb bible, drinking the juice of the leaves or roots of burdock with old wine could "wonderfully help the biting of serpents".

If bitten by a "mad dog", a salve made from burdock root beaten with a little salt, placed directly on the wound would instantly ease the pain.

Parsley Resides with the Devil

There are many superstitions attached to parsley.

Taking as long as six weeks to germinate, parsley was often linked to the Devil. Superstition tells us that its roots stretched to hell and back nine times before it sprouted.

Parsley seeds should be sown on Good Friday around 3 pm – the time of Jesus' crucifixion – because the Devil "is powerless and preoccupied" during this period. If the seeds never sprout, the Devil lies beneath the soil.

Garlic Triumphs Over Evil

Folklore surrounding garlic suggests that it had the power to ward off evil, including, of course, being the bane of every vampire's existence.

Sailors packed garlic for eventful voyages and Spanish matadors wore necklaces made with its cloves in the bullring.

It was believed to be the nemesis of witches and wild garlic was hung outside homes on wreathes, stuffed into keyholes or rubbed on chimneys.

Paradoxically, others believed that eating a clove of garlic on an empty stomach would bring good luck.

The Divine Farmer

The Chinese have used herbal medicine for centuries. The revered father of Chinese herbalism is mythical emperor Shennong, also known as the "Divine Farmer". He is said to have taught his people to plough the land, grow crops and introduced basic agricultural methods to the region.

Ancient texts tell us that he tasted hundreds of herbs for use in healing and medicine. Now considered a folklore hero in China and Vietnam, he is thought to have died from eating a poisonous plant.

Myrtle for Love

The word "myrtle" derives from the Greek, myrtos.

To the ancient Greeks, myrtle was the herb of love, associated with Aphrodite (the goddess of love, beauty and pleasure). They would wear garlands of myrtle to encourage new love and prosperity.

In Roman legend, the goddess Venus wore a wreath of myrtle when she was handed the golden apple of Paris in recognition of her beauty. She also held a sprig of myrtle as she rose from the sea, and was said to have turned a priestess into myrtle to protect her from an overeager suitor.

Chapter Three

AN A-Z OF MEDICINAL HERBS

Aloe Vera

Aloe barbadensis

Also known as: aloes

PLANT FAMILY:
Asphodelaceae

COMMONLY USED FOR:
Healing, skincare, sunburn, ulcers

CAUTIONS:
Do not use bitter aloes on skin.
Not to be taken internally
if pregnant or breastfeeding.

This prickly succulent is native to Africa, but has been cultivated worldwide thanks to its remarkable effectiveness in healing cuts and burns.

It has two types of medicinal use. The first (and best-known) is the clear gel in the leaves, which coats wounds and stimulates the immune system, speeding up healing.

The second is the yellow sap (bitter aloes) at the base of the leaf. This contains anthraquinones, which are strong laxatives for short-term constipation.

Aloe vera is also easy to propagate at home: break off small, rooted plantlets and repot. To harvest, split the leaves open to collect the gel or bitter aloes.

Uses & Preparation of Aloe Vera

CONSUME
Aloe vera juice is an everyday supplement used for heartburn, cholesterol and blood sugar. Mix around 2 tbsp of juice with water and drink.

TOPICAL
Use ointment for burns and skin conditions. Apply the gel to the site twice a day.

Ashwagandha

Withania somnifera

Also known as: Indian ginseng

PLANT FAMILY:
Solanaceae

COMMONLY USED FOR:
Stress, sleep problems

The ideal antidote to busy modern life, ashwagandha is used for restfulness and improving sleep quality.

The effects are often linked to naturally occurring compounds called withanolides that help to decrease anxiety and lower stress levels in the body.

The evergreen shrub grows in India and is commonly compared to the herb ginseng, which is used to improve vitality.

Uses & Preparation of Ashwagandha

CAPSULES
Powdered root capsules can be taken for exhaustion.

CONSUME
The berries (fresh or dried) can be chewed to help with recovery after surgery.

DRINK
½ tsp powder made from the leaves and dissolved in a little water is recommended for anemia. Suitable for children and the elderly to support the immune system.

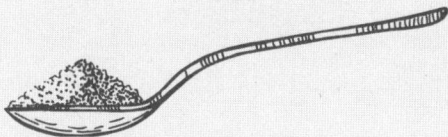

Basil

Ocimum basilicum

Also known as: sweet basil

PLANT FAMILY:
Lamiaceae

COMMONLY USED FOR:
Digestive problems

CAUTIONS:
Sweet basil oil should not
be taken internally.

Perhaps one of the most common herbs, basil is a culinary staple.

Its medicinal uses relate to how the body breaks down food. Taken internally, it acts on the upper digestive tract and nervous system to ease cramps, nausea and vomiting.

Since Roman times, the herb has been eaten to relieve flatulence and as a diuretic to flush out poisons.

> *"The smell of Basil is good for the heart, takes away sorrowfulness and makes a man merry and glad."*

John Gerard,
Herball, 1597

Uses & Preparation
of Basil

BATHE

Mix 6–12 drops essential oil with 2 tbsp carrier oil like jojoba to make sweet basil oil. Add to baths for muscle relief.

CONSUME

Eating basil leaves regularly can be beneficial for the digestive system.

TOPICAL

Since the times of the ancient Greeks, the juice from sweet basil leaves has been used to relieve pain from stings and bites. Rubbing leaves on skin acts as an insect repellent.

Borage

Borago officinalis

Also known as: starflower, burrage, bee bread

PLANT FAMILY:
Boraginaceae

COMMONLY USED FOR:
Respiratory problems, skin conditions

CAUTIONS:
Seek professional advice before taking borage due to toxic pyrrolizidine alkaloids. Do not take internally.

Since Roman times, borage has been used as a symbol of mental strength. The Celtic word "*borrach*" means "courage".

In his *Herball*, Gerard wrote: "I, borage, bring always courage."

Soldiers are said to have taken borage wine before a battle and we now know that the herb supports the adrenal glands, which produce stress hormones. Targeting these glands could explain why it is effective in treating fevers, too.

Borage honey is produced from honeybees pollinating the borage plant.

Uses & Preparation of Borage

CONSUME
Borage flowers increase sweating and lower body temperature during fevers.
Borage honey has antiseptic, anti-inflammatory and antioxidant properties.

DRINK
Infuse 2 tsp dried flowers in 1¾ pints (1 litre) boiled water to create a tea that treats coughs, colds and fevers.

TOPICAL
Apply a cold compress with borage leaves to treat burns and stings.

Burdock

Arctium lappa

Also known as: niu bang zi, peronata,
bardona, clotbur

PLANT FAMILY:
Asteraceae

COMMONLY USED FOR:
Throat infections, rashes, skin conditions

CAUTIONS:
Can cause dermatitis.

Burdock is sometimes known as a weed because it is so common worldwide, but it is revered as one of the foremost detoxifying herbs.

The seeds and dried root help to flush toxins out of the body thanks to cleansing and diuretic properties.

Fresh roots are also thought to have an antibiotic effect because they contain an active compound called polyacetylenes.

Uses & Preparation of Burdock

DRINK
Simmer 2 tsp burdock root and 5 tsp dandelion root in boiling water to create an infusion. Strain and store in a cool place. Drink 5 fl oz (150 ml) twice a day.

TINCTURE
Dissolve burdock root in 35–40 per cent vodka (1:5 ratio) over 14 days. Take 20 drops diluted with water 2–3 times a day for skin disorders.

Calendula

Calendula officinalis

Also known as: marigold

PLANT FAMILY:
Asteraceae

COMMONLY USED FOR:
Minor burns, nappy rash, inflammation

CAUTIONS:
Can cause an allergic reaction.

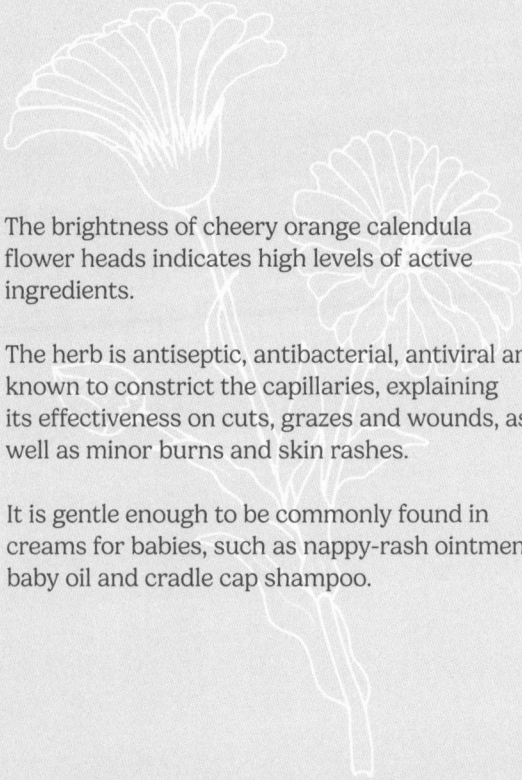

The brightness of cheery orange calendula flower heads indicates high levels of active ingredients.

The herb is antiseptic, antibacterial, antiviral and known to constrict the capillaries, explaining its effectiveness on cuts, grazes and wounds, as well as minor burns and skin rashes.

It is gentle enough to be commonly found in creams for babies, such as nappy-rash ointment, baby oil and cradle cap shampoo.

Uses & Preparation of Calendula

BATHE
Combine the juice of burdock leaves with vinegar and add to a hot bath to reduce swelling.

DRINK
Steep 1–2 tbsp dried or fresh calendula flowers in 8 fl oz (240 ml) simmering water for 15–20 minutes. Strain and enjoy.

TOPICAL
Massage essential oil onto inflamed skin 2–3 times a day. Apply calendula ointment 3 times a day for minor burns.

Cayenne

Capsicum annuum

Also known as: chilli

PLANT FAMILY:
Solanaceae

COMMONLY USED FOR:
Headaches, arthritis, indigestion

CAUTIONS:
Use caution when handling raw chillies.

Native to tropical regions of the Americas, like Mexico, cayenne only came to Europe in the 16th century.

We know the fruit best for its hot, spicy taste, but within the body, cayenne can also fire up circulation by improving blood flow and easing conditions like gastroenteritis and diarrhoea.

On the surface of the skin, the compound capsaicin desensitizes nerve endings for effective pain relief.

Uses & Preparation of Cayenne

CAPSULES
Take 1–2 tablets daily for poor circulation.

GARGLE
Dilute 2 tsp powdered cayenne with 2 tbsp each of lemon juice, water and honey. Gargle for a sore throat.

TOPICAL
Create an infused oil by gently simmering 3½ oz (100 g) chopped chilli with 16 fl oz (500 ml) oil. Allow to cool then massage onto muscles to relieve soreness and inflammation.

Chamomile

Matricaria chamomilla

Also known as: English chamomile, garden chamomile, mother's daisy, wild chamomile

PLANT FAMILY:
Asteraceae

COMMONLY USED FOR:
Gastritis, Crohn's disease, irritable bowel syndrome

CAUTIONS:
Avoid essential oil in pregnancy or with blood thinners.

Best known as a herbal tea, chamomile comes from a daisy-like flower seen throughout Europe.

As an infusion, chamomile is kind to the digestive system and is known to relax tension in the gut.

Conditions such as gastritis, Crohn's disease and irritable bowel syndrome have been treated using chamomile. It is so gentle, yet effective, that it can be used on infants, especially those with colic.

Uses & Preparation of Chamomile

DRINK
Infuse flower heads in a pot of boiled water. Simmer for 10 minutes. Strain and then drink 5 fl oz (150 ml) before bed.

TOPICAL
Used externally, chamomile ointment can be used to soothe sore, itchy skin.

Combine 4 drops essential oil with 1 tbsp coconut oil for nappy rash.

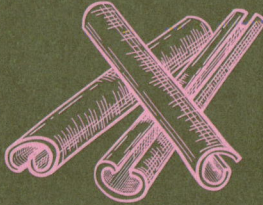

Cinnamon

Cinnamomum spp

Also known as: dalcini, rou gui

PLANT FAMILY:
Lauraceae

COMMONLY USED FOR:
Colds, flu, digestive issues, high blood sugar

CAUTIONS:
Do not take essential oil internally.
Overuse could lead to low blood sugar.

Since ancient times, cinnamon has had an exotic and sought-after appeal. It is one of the oldest known spices, with a history dating back to 2700 BCE.

Cinnamon trees are native to India and Sri Lanka, but the spice didn't reach Europe until 500 BCE.

Traditionally, it was used to treat common colds and flu; now it is also used to aid digestion and metabolizing sugars.

Uses & Preparation of Cinnamon

DRINK
Add 1 tsp ground powder to 8 fl oz (240 ml) boiled water. Drink 2½ fl oz (75 ml) 2–3 times a day to treat cold and flu symptoms.

CONSUME
Sprinkle ground cinnamon over porridge or toast to regulate blood sugar.

TINCTURE
Infuse the herb with 35–40 per cent vodka (1:5 ratio) over 14 days. Use a wine press to extract the liquid. Take 20 drops with water 4 times a day to treat flatulence.

Dandelion

Taraxacum officinale

Also known as: lion's teeth, pissenlit, fairy clock, canker weed, priest's crown, blowclock

PLANT FAMILY:
Asteraceae

COMMONLY USED FOR:
Digestion, liver problems, eczema

Instantly recognizable for their tufted yellow flower heads, dandelions are so prevalent in Europe that, to many gardeners, they are weeds.

In herbal medicine, however, they are a wildflower with a difference.

The leaves are high in potassium and act as an effective diuretic for fluid retention, while the root is a superb prebiotic and detoxifying herb.

Uses & Preparation of Dandelion

CAPSULES

Take diuretic dandelion tablets to counteract water retention.

CONSUME

Young and tender leaves can be used to flavour salads.

DRINK

Take a capful of dandelion juice 3 times daily before a meal to ease fluid retention.

Combine the leaves with boiled water. Simmer and then strain after 10 minutes. Take 16 fl oz (500 ml) once a day for acne.

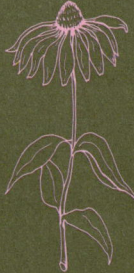

Echinacea

Echinacea spp.

Also known as: purple coneflower, Sampson
root, black Sampson, Indian snakeroot

PLANT FAMILY:
Asteraceae

COMMONLY USED FOR:
Infections, colds, flu

CAUTIONS:
Can cause allergic reactions.

Long before modern medical research, indigenous tribes used the bold purple plant echinacea to treat snake bites, which explains the nickname "Indian snakeroot".

The roots are thought to stimulate the immune system to more effectively fight off viral and fungal infections. This is thanks to its content of polysaccharides (which inhibit viruses from taking over the cells) and alkylamides (which are antibacterial and antifungal).

Uses & Preparation of Echinacea

CAPSULES
Powdered root capsules can be taken to treat the common cold.

GARGLE
Add roots to a pot of cold water and bring to the boil to create a decoction. Simmer for 30 minutes, then strain. Gargle 3½ tbsp (50 ml) 3 times daily for throat infections.

TINCTURE
Infuse roots with 35–40 per cent vodka (1:5 ratio) over 14 days. Use a wine press to extract the liquid. Take ½ tsp in water 3 times a day for chronic infections.

Evening Primrose

Oenothera biennis

Also known as: evening star, sundrop,
German rampion, hogweed,
king's cure-all, fever-plant

PLANT FAMILY:
Onagraceae

COMMONLY USED FOR:
Digestive problems, asthma, bloating

CAUTIONS:
Not suitable for people with epilepsy.

The nickname "king's cure-all" gives a clue to the history of evening primrose as an all-rounder in 17th-century Europe.

It has well-known applications in everything from sore throats to hemorrhoids.

The flowers, leaves and stem bark are all useful, but oil from the seeds is known to have the most benefits because it contains essential, naturally occurring omega-6 fatty acids.

Uses & Preparation of Evening Primrose

CONSUME

Take ¼ tsp essential oil per day to help maintain healthy blood pressure. Widely available as a supplement.

TOPICAL

Apply the essential oil to the skin for eczema.

Make a poultice by crushing herbs into a pulp using a pestle and mortar, adding just enough water to make a paste. Spread on a cloth and apply directly to bruises or wounds.

Feverfew

Tanacetum parthenium

Also known as: bachelor's buttons, featherfew, medieval aspirin

PLANT FAMILY:
Asteraceae

COMMONLY USED FOR:
Fevers, migraines, arthritis

CAUTIONS:
Do not take in pregnancy or in combination with blood thinners. Can cause allergic reaction. Fresh leaves can cause mouth ulcers.

As the name would suggest, feverfew has traditional uses in cooling the body and treating a temperature. In fact, after proving its efficacy with headaches, it was long referred to as "medieval aspirin".

It is still predominantly used to prevent migraines and treat headaches associated with menstruation.

Often it is recommended as a long-term treatment and will be taken as capsules or tablets.

Uses & Preparation of Feverfew

CAPSULES
Tablets provide relief for headaches
and migraines.

CONSUME
Eat fresh leaves 2–3 times a day for migraines.

DRINK
Steep herbs in boiling water for 15 minutes
to make feverfew tea. Strain and sweeten with
honey for flavour.

TINCTURE
Place fresh leaves and flowerheads in 35–40
per cent vodka (1:5 ratio) for 3 weeks. Strain
the mixture into a clean jar. Take 10 drops a day
to prevent migraines.

Flaxseed

Linum usitatissimum

Also known as: flax, linseed

PLANT FAMILY:
Linaceae

COMMONLY USED FOR:
Diabetes, constipation, high cholesterol,
heart disease, cancer

CAUTIONS:
Immature seeds could be toxic. Store
cracked or ground seeds in an airtight
container in the fridge.

With a Latin name that literally translates to "very useful", you can't go far wrong with flaxseed in your herbal healing kit.

It's been harvested since the beginning of civilization for making clothes and is only now being hailed as the new superfood.

Flaxseed contains antioxidants that help to prevent disease by removing damaging free radicals (unstable molecules) from the body.

Uses & Preparation of Flaxseed

CAPSULES
Take flaxseed capsules for digestive health and to help control blood sugar.

CONSUME
Sprinkle ground flaxseed on smoothies, yogurt or breakfast cereal.

Drizzle flaxseed oil over salads for a healthy alternative to dressing.

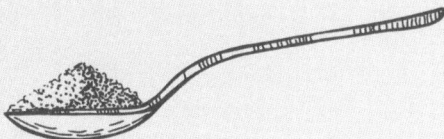

Garlic

Allium sativum

Also known as: bulbus, nectar of the Gods, poor man's treacle, stinking rose

PLANT FAMILY:
Liliaceae

COMMONLY USED FOR:
Infections, circulatory issues, cookery, boosting the immune system

CAUTIONS:
Not for use if under 12 years old. Seek medical advice if taking blood-thinning medication.

The bulbous perennial garlic has been studied extensively for its medicinal properties.

Findings show that it can be effective in circulatory problems: normalizing blood fat levels, lowering blood pressure, keeping blood thin and protecting against blood clots.

As far as at-home uses go, garlic is often used to treat common infections of the nose, throat and chest. It is also strongly antifungal.

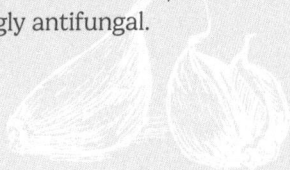

Uses & Preparation of Garlic

CAPSULES
Take dried garlic capsules for multiple health benefits.

CONSUME
Include chopped raw garlic in your diet to lower blood pressure.

Crush garlic to release allicin, which is known to help fight infections.

DRINK
Steep a clove of garlic in a pot of water until boiling. Simmer for 5 minutes, then strain to drink as a tea.

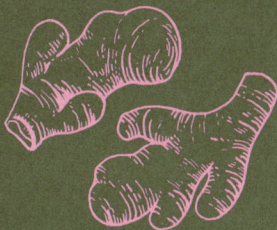

Ginger

Zingiber officinale

Also known as: zingiber

PLANT FAMILY:
Zingiberaceae

COMMONLY USED FOR:
Morning sickness, digestive issues,
circulation, pain relief

CAUTIONS:
Limit dosage to scant 1 tsp (4 g) fresh root
per day in pregnancy. Do not take essential
oils internally. Avoid medicinal doses if
suffering from peptic ulcers.

As well as being a widely used spice and food flavouring, ginger (from the rhizome of the perennial) is revered as one of the world's best herbal medicines, often used to treat a string of digestive problems including: wind, bloating, cramps and indigestion.

It has known effectiveness in treating morning sickness in the early stages of pregnancy and relieving general feelings of nausea.

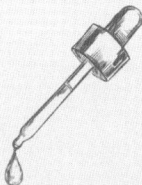

Uses & Preparation of Ginger

CAPSULES
Take a supplement for morning sickness.

DRINK
Simmer fresh ginger in boiled water for 10 minutes then strain. Drink 5 fl oz (150 ml) 3 times a day for nausea.

TINCTURE
Dissolve fresh ginger in 35–40 per cent vodka (1:5 ratio) over 14 days. Use a wine press to extract the liquid. Take 30 drops with water twice a day.

TOPICAL
To improve circulation, rub 1–2 drops of oil over the heart twice a day.

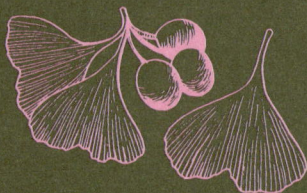

Ginkgo

Ginkgo biloba

Also known as: maidenhair tree, bai guo

PLANT FAMILY:
Ginkgoaceae

COMMONLY USED FOR:
Circulation, dementia

CAUTIONS:
Raw ginkgo seeds are poisonous.

Ginkgo is a deciduous tree first thought to have grown in China 190 million years ago. It is therefore widely used in Traditional Chinese Medicine (TCM).

The seeds are used to open the organ channels, improving blood flow to the kidneys, liver, brain and lungs. Ginkgo extract, which is used more in Western medicine, comes from the fan-shaped leaves.

It is an antioxidant, and is antiallergenic and anti-inflammatory.

Uses & Preparation of Ginkgo

CAPSULES
Take one tablet daily to improve blood flow and aid memory loss.

DRINK
Take 10 drops fluid extract 2–3 times a day in a small amount of water.

TINCTURE
Infuse the leaves with 35–40 per cent vodka (1:5 ratio) over 14 days. Use a wine press to extract the liquid. Take 1 tsp with water 2–3 times daily.

Ginseng

Panax ginseng

Also known as: ren shen

PLANT FAMILY:
Araliaceae

COMMONLY USED FOR:
Inflammation, brain function

CAUTIONS:
Avoid during pregnancy. Do not take
in combination with caffeine.
Do not exceed the dosage.

The Chinese herb ginseng has a long and powerful history. Around 7,000 years ago, it was so sought after that wars were fought over the forests in which it grew.

The herb is hailed for building immunity, regulating blood sugar, improving focus and reducing inflammation.

Over the years, it has been overharvested and is now rare to find in the wild. It takes between 5–10 years to grow so it's close to being endangered.

Uses & Preparation of Ginseng

CAPSULES
Take daily with food for fatigue.

DRINK
Steep fresh roots in a pot of boiled water.
Simmer for 5 minutes then strain and drink.

CONSUME
Add root extract powder to smoothies for
improved energy.

Sprinkle a pinch of dried root into
vegetable soup.

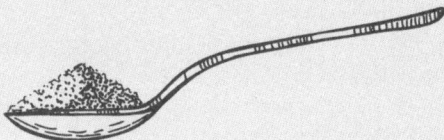

Goldenseal

Hydrastis canadensis

Also known as: orangeroot, yellow puccoon

PLANT FAMILY:
Ranunculaceae

COMMONLY USED FOR:
Infections, stomach and liver problems

CAUTIONS:
Toxic if taken in excess. Avoid in pregnancy
and breastfeeding. Not suitable for children.
Do not take with high blood pressure.

Once abundant in North American forests, the herbaceous perennial goldenseal is now an endangered species.

Until 1997, it was excessively harvested in the wild owing to its reputation as a catch-all remedy.

Goldenseal was principally used by the Cherokee as an insect repellent and for treating infections. Later, it was exported to Europe by the tonne to treat colds, wounds and digestive problems.

Uses & Preparation of Golden Seal

CAPSULES
Take capsules of powdered root daily with food.

DRINK
Take 14–28 drops liquid extract (1 ml) 3 times a day in a small amount of water. Use over 10 days.

Steep 2 tsp of the dried herb in 8 fl oz (240 ml) boiled water for 15 minutes to drink in a tea.

Lavender

Lavandula angustifolia syn.

Also known as: English lavender, common lavender, elf

PLANT FAMILY:
Lamiaceae

COMMONLY USED FOR:
Depression, indigestion, wind, bloating, wounds, burns, sores, asthma, muscle tension

CAUTIONS:
Do not take lavender essential oil internally.

Lavender oil didn't gain widespread popularity for its medicinal qualities until the early 20th century when it was used to dress wounds during World War I to replace the dwindling supplies of antiseptics.

It is believed that housewives and Girl Guides were tasked with collecting lavender to aid the effort.

Its uses are widespread today, but it still has its best applications in first aid.

Uses & Preparation of Lavender

DRINK
Enjoy lavender tea regularly to boost the immune system.

TINCTURE
Add 4 oz (115 g) lavender flowers to 16 fl oz (500 ml) vodka in a glass jar. Seal and shake, then leave for at least 14 days. Strain using a coffee filter. Dilute in water and take for stress.

TOPICAL
Mix 3–4 drops essential oil with 1 tbsp coconut oil and apply to wounds with a cotton ball.

Milk Thistle

Silybum marianum syn.

Also known as: Mary thistle, marian thistle,
St Mary's thistle, our lady's thistle

PLANT FAMILY:
Asteraceae

COMMONLY USED FOR:
Liver conditions, depression.

In a quote from Gerard's *Herball*, it was said: "My opinion is that [milk thistle] is the best remedy that grows against all melancholy diseases." Quite the endorsement.

Recent research proves that this spiny biennial could have an impact on depression via its protective effect on the liver.

It's the silymarin – a substance contained within the seeds – that's thought to have healing properties.

Uses & Preparation
of Milk Thistle

CAPSULES
Take 2 capsules with food to treat a hangover.

DRINK
Make a decoction by adding seeds to a pot of
cold water and bring to the boil. Simmer for 30
minutes then strain. Take 2½ fl oz (75 ml) a day
for liver infections.

CONSUME
Sprinkle raw seeds on food or grind them into a
powder and add to smoothies.

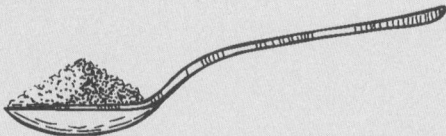

Mugwort

Artemisia vulgaris

Also known as: mater herbarum, moxa, sailor's tobacco, mugger, muggart, felon herb

PLANT FAMILY:
Asteraceae

COMMONLY USED FOR:
Menstrual cramps, digestion

CAUTIONS:
Do not use during pregnancy. Mugwort can interfere with certain allergies.

From the name of it, the roadside perennial mugwort could be straight from the pages of Harry Potter.

During the Middle Ages, it was believed to possess magical powers as garlands or belts made of the plant were supposed to protect from evil spirits.

Later, it became known as mater herbarum, or the "mother of herbs", because it was used to ease menstrual cramps and help labour.

Uses & Preparation of Mugwort

BURN
Mugwort is sometimes burned and used for aromatherapy in a method called moxibustion.

CONSUME
Add aromatic mugwort to soups or lamb stews, or sprinkle young stems over salads.

DRINK
Steep mugwort leaves in boiling water for 10 minutes then strain. Drink 2–3 cups a day.

TINCTURE
Take up to 40 drops (2 ml) 3 times a day in a little water.

Nettle

Urtica Dioica

Also known as: stinging nettle, burning
weed, ettle, heg-beg, hop-tops

PLANT FAMILY:
Urticaceae

COMMONLY USED FOR:
Arthritis, anemia, hay fever

Nettles are one of the first plants we learn to recognize and avoid due to their nasty stings on contact with the skin. However, for conditions like arthritis, the leaves can have a counterirritant effect that overrides musculoskeletal pain.

The root, which is high in flavonoids and potassium, is also a diuretic used for cleansing and detoxifying. It is also anti-inflammatory, which helps with nasal blockages in hay fever.

Uses & Preparation of Nettle

CAPSULES
Dried nettle leaf capsules are used for urinary support. Take 2 a day.

CONSUME
Add 14 oz (400 g) nettle leaves to carrot, leak, potato and onion to make a soup that's rich in iron.

DRINK
Drink nettle tea 3 times a day for arthritis.

TOPICAL
Nettle creams and ointments can be rubbed into the skin to relieve arthritis pain and inflammation.

Oregano

Origanum vulgare

Also known as: wild marjoram,
joy of the mountains

PLANT FAMILY:
Lamiaceae

COMMONLY USED FOR:
Gastroenteritis, bronchitis, tonsillitis

CAUTIONS:
Do not use during pregnancy.

From the kitchen cupboard, oregano is a staple herb in the Western diet.

Not only does it have a sweet taste and distinctive aroma, it has been found to contain bacteria-fighting ingredients and is therefore helpful in treating infections (such as tonsillitis and bronchitis). It has also been shown to reduce inflammation because it is rich in antioxidants.

"Sweet Marjoram is good for those who are given to over-much sighing."

John Gerard,
Herball, 1597

Uses & Preparation of Oregano

DRINK

For coughs and bronchitis, add a sprig of oregano to boiled water for a few minutes then strain. Sweeten with honey.

CONSUME

Make oregano oil for cooking using 2 tbsp dried oregano and 12 fl oz (350 ml) olive oil.

Use fresh or dried oregano for a variety of recipes including salads, pesto, sauces and meat dishes.

Peppermint

Mentha x piperita

Also known as: mint

PLANT FAMILY:
Lamiaceae

COMMONLY USED FOR:
Wind, bloating, flatulence, colic

CAUTIONS:
Not suitable for children under 12.

Peppermint is a strongly aromatic plant grown all over the world. Its leaves can soothe a variety of digestive problems thanks to an "antispasmodic" effect, which eases complaints such as nausea, bloating and flatulence.

It can also soothe the lining of the colon to relieve diarrhoea. On the surface of the skin, volatile oil contained within the leaves shows antiseptic, antifungal and anesthetic properties.

"The savor of the water mint rejoiceth the heart of men."

John Gerard,
Herball, 1597

Uses & Preparation of Peppermint

CAPSULES
Take peppermint oil capsules around 1 hour before meals for indigestion.

DRINK
Steep 1 tsp dried peppermint leaves in 8 fl oz (240 ml) boiling water for 10 minutes. Strain, cool and drink.

TINCTURE
Place a large bunch of peppermint leaves in a jar, and fill with 35–40 per cent vodka. Leave for 1 month then strain. Take 20 drops up to 3 times a day to help with digestive issues.

Rosemary

Salvia rosmarinus

Also known as: dew of the sea

PLANT FAMILY:
Lamiaceae

COMMONLY USED FOR:
Memory, low blood pressure, migraines

CAUTIONS:
Do not take internally.

Rosemary has a long history of being used in herbal remedies and folklore; in Elizabethan England, wedding couples wore a sprig of rosemary to symbolize fidelity.

But there is a lot more to this herb. It has multiple benefits, beyond its distinct and powerful flavour on food.

In fact, rosemary has invigorating properties used to aid memory and blood flow to the brain, as well as a power to improve migraines or headaches. Scientists note the active constituents of rosmarinic acid and carnosol, thought to contribute to brain function.

Uses & Preparation of Rosemary

BURN
Place several drops of rosemary oil in an oil burner to improve concentration.

DRINK
Place ½ sprig of rosemary in 8 fl oz (240 ml) boiled water and drink to boost memory.

Create a pot of infused rosemary by adding ¾ oz (20 g) dried rosemary to 5 fl oz (150 ml) just-boiled water. Take 2 tbsp every 3 hours for a headache.

Sage

Salvia officinalis

Also known as: common sage, garden sage, kitchen sage, broadleaf sage, Dalmatian sage

PLANT FAMILY:
Lamiaceae

COMMONLY USED FOR:
Sore throats, poor digestion, memory, attention, menopause

CAUTIONS:
Do not take as a herbal remedy during pregnancy.

Its Latin name, *Salvia*, means "to cure" and it has long been highly regarded for its medicinal value.

Sage is high in magnesium, zinc and copper, and also contains vitamins A, C, E and K.

It is known to have over 160 types of polyphenols, which are anti-inflammatories and antioxidants. As such, it has been used to improve everything from a sore throat to memory, processing and attention.

Uses & Preparation of Sage

GARGLE
Gargle an infusion of fresh leaves and boiled water (strained and cooled) up to 3 times a day for sore throats.

TINCTURE
Dissolve 1-part herb in 5 parts alcohol over 14 days. Use a wine press to extract the liquid. Take 40 drops (2 ml) twice a day with water.

TOPICAL
Rub sage leaves on stings and bites as a first-aid remedy.

St John's Wort

Hypericum perforatum

Also known as: SJW, klamath weed, tipton weed, goat weed, enola weed, amber

PLANT FAMILY:
Hypericaceae

COMMONLY USED FOR:
Depression, menopause

CAUTIONS:
Do not eat the plant directly. Do not combine with other medicines.

It used to be called a cure for "all down-heartedness".

Even the sight of its sunshine-yellow clusters was traditionally enough to brighten your mood.

Still today, St John's wort is best associated with treating ailments of the mind like depression or seasonal affective disorder because it is thought to act on neurotransmitters in the brain like serotonin.

Uses & Preparation of St John's Wort

CAPSULES
Take daily with water for mood regulation.

TINCTURE
Fill an airtight jar two thirds full with fresh flowering tops. Cover with 35–40 per cent vodka and seal. Shake every few days for 14 days. Take 20 drops with water 3 times a day.

Tea Tree

Melaeuca alternifolia

Also known as: melaeuca oil

PLANT FAMILY:
Myrtaceae

COMMONLY USED FOR:
Athlete's foot, skin infections, acne, stings,
wounds, burns

CAUTIONS:
Do not take essential oil internally.

Native to Australia, the leaves and flowers of
tea tree are used to make a world-renowned
essential oil revered for its antiseptic,
antibacterial, antifungal and antiviral properties.

It is typically made into creams and drops and
applied directly to the skin to treat wounds,
stings and burns as a first-aid remedy or used
as a longer course to improve skin conditions
like acne.

Uses & Preparation of Tea Tree

TOPICAL

Add a few drops of essential oil to 1-part water and 1-part rubbing alcohol for a useful spray to treat fungal infections such as athlete's foot.

Steep the leaves in boiled water and allow to cool. Use on the scalp to treat dandruff.

Combine a base cream ointment with 5 drops essential oil for nail fungus and insect bites.

Turmeric

Curcuma longa syn.

Also known as: haldi, jiang huang

PLANT FAMILY:
Zingiberaceae

COMMONLY USED FOR:
Inflammation, arthritis, diabetes, psoriasis

CAUTIONS:
If used in combination with blood-thinning
medication, consult a professional.

Commonly known for curries, food colouring and the kind of stubborn staining that could ruin your clothes, turmeric is undergoing a significant resurgence as a healing herb like no other.

Since the early 2000s, turmeric has been at the centre of many medical studies with outcomes showing that it can block inflammatory pathways and therefore make an effective treatment in arthritis and in lowering cholesterol.

Uses & Preparation of Turmeric

CAPSULES
Capsules with black pepper improve absorption of turmeric and therefore effectiveness.

CONSUME
Add turmeric powder to curries, milk or scrambled eggs, or sprinkle on soups to maintain healthy blood sugar levels.

DRINK
Add ½ tsp dried turmeric powder to a cup of boiled water (or plant milk for a latte). Add lemon juice and honey for flavouring.

Valerian

Valeriana officinalis

Also known as: setwall, phu

PLANT FAMILY:
Valerianaceae

COMMONLY USED FOR:
Stress, anxiety, muscle tension

CAUTIONS:
Can cause drowsiness. Not to be used in
combination with sleep aids.

Valerian relaxes the nervous system and helps with conditions like anxiety and stress.

In the past, it has been used to treat epilepsy – even as far back as 1592 – but today it is better associated with improving sleep and clearing an overactive mind.

It is used to relieve period pains, muscle spasms and tension in the body. It has also been shown to reduce blood pressure.

Uses & Preparation of Valerian

CAPSULES

Take daily to treat insomnia.

DRINK

For sleep, pour 8 fl oz (240 ml) boiling water over 1 tsp dried root. Brew for 5–10 minutes then strain and enjoy around 1 hour before bedtime.

Go Forth and Forage!

There is so much to gain when we take the time to forge a closer bond with nature, and discover the inherent value that plants hold.

As you embrace the opportunity to use herbs medicinally and for your wellbeing, take time to also get outside and connect to the nature that surrounds you – grounding yourself in green spaces, forests and woodlands, and gardening or foraging for herbs.

"Look deep into nature, and then you will understand everything better.„

Albert Einstein (1879–1955)